Astrobiologist

WEIRD
CAREERS
in SCIENCE

Astrobiologist

Mary Firestone

CHELSEA HOUSE
PUBLISHERS
A Haights Cross Communications Company ®
Philadelphia

CHELSEA HOUSE PUBLISHERS

VP, NEW PRODUCT DEVELOPMENT Sally Cheney
DIRECTOR OF PRODUCTION Kim Shinners
CREATIVE MANAGER Takeshi Takahashi
MANUFACTURING MANAGER Diann Grasse
SERIES DESIGNER Takeshi Takahashi
COVER DESIGNER Takeshi Takahashi

STAFF FOR **ASTROBIOLOGIST**

PROJECT MANAGEMENT Ladybug Editorial and Design
DEVELOPMENT EDITOR Tara Koellhoffer
LAYOUT Gary Koellhoffer

A Haights Cross Communications ✦ Company ®

www.chelseahouse.com

First Printing

9 8 7 6 5 4 3 2 1

Library of Congress Cataloging-in-Publication Data

Firestone, Mary.
  Astrobiologist / Mary Firestone.
     p. cm. — (Weird careers in science)
  Includes bibliographical references.
  ISBN 0-7910-8971-1
  1. Space biology—Juvenile literature. 2. Exobiology—Juvenile litera-
ture. 3. Space biology—Vocational guidance—Juvenile literature. 4.
Exobiology—Vocational guidance—Juvenile literature. I. Title. II. Series.
  QH327.F545 2005
  576.8'39—dc22
                                              2005017658

TABLE OF CONTENTS

Chapter 1

Introduction

ASTROBIOLOGIST KEVIN HAND has climbed inside a tiny **submersible**, which is about to lower him slowly down through the waters of the Pacific. When he finally reaches the ocean floor, he'll get a close look at some strange ocean organisms that actually love living next to the smelly, superheated, smoky vents in the ocean floor called "**black smokers**" (Figure 1.1). Hand is exploring these areas as part of the research team for the IMAX film *Aliens of the Deep*.

Black smokers look like thick tubes with bumpy sides. Dense clouds of iron, copper, and hydrogen sulfide flow out of them. In an article about Kevin Hand's work on this film, a writer for

Figure 1.1 Scientists have recently discovered strange new life forms that not only survive but thrive near black smokers on the ocean floor, like the one seen here.

the *Stanford (University) Report* describes groups of black smokers as an "upside-down power plant where microbes eat scalding exhaust." And that's exactly what many organisms do.

Until recently, scientists didn't know life could exist in the extreme environment of the ocean, where there is no light and where temperatures reach a blistering hot 345°C (653°F). But the **microbes** that live there are like no others. They are newly discovered organisms that scientists are just finding out about, called "**extremophiles**."

To help film *Aliens of the Deep*, Hand worked with other astrobiologists, marine biologists, planetary scientists, and geophysicists from the National Aeronautics and Space Administration (NASA). He completed eight different dives into the Pacific and Atlantic oceans in the tiny submersible, to visit smoker sites with names like "Lost City" and "Snake Pit." He spent up to 15 hours at a time under the sea, where the film crew recorded images of 6-foot-long tubeworms, blind white crabs, and massive quantities of heat-sensing white shrimp.

But why, you may ask, are **astrobiologists** here? What do these strange undersea creatures have to do with life in outer space? Astrobiologists are very interested in this project, because extreme conditions near **hydrothermal vents** may resemble life on other planets and moons. The weather in outer space is extreme. If organisms can survive and thrive in similar extreme conditions here on Earth, then maybe similar life-forms exist in outer space.

Kevin Hand says of the extremophile discoveries, "This perhaps reflects the most important lesson learned from the discovery of the vents back in the late 1970s." It has "caused the biological community to scratch its head and rethink things [about places where life might exist]."

WHAT IS ASTROBIOLOGY?

Astrobiology is the study of life in the universe. It includes the fields of astronomy, physics, biology, geology, paleontology, and many others, such as microbiology. The curiosity of scientists who wonder if life exists somewhere besides Earth has created this new science. Since we cannot yet travel to other planets, astrobiologists must begin with the only life we know about: life on Earth.

The recent discovery that there is a whole chain of life thriving on the heat and chemicals boiling up out of the bottom of the sea floor has caused a big leap in research activity for astrobiology (Figure 1.2). These hot spots are energizing long food chains, which begin with little microbes that eat hydrogen sulfide from the black smokers. Then other bigger creatures come and eat the microbes, and the chain of life continues, with little fish being eaten by bigger fish.

In an interview with PBS for his book *The Universe Below: Discovering the Secrets of the Deep*, *New York Times* science writer William Broad described the vast area now being explored by astrobiology: "The **biosphere** is the habitable part of the Earth. If you take all the stuff we have on land—the trees, all the stuff we know and love, the

Black Smoker

A strange type of shrimp was discovered near black smokers. Called *Rimicaris exoculata*, this shrimp has eyes on its back! Biologist Cindy Van Dover made the discovery while she was exploring the sea floor in a submersible. The submersible was used to collect samples from the areas near black smokers, to be analyzed later by scientists.

Figure 1.2 Astrobiologists study life on the ocean floor because it may closely resemble conditions in which life may exist on other planets. Coral-like formations like these found in Cuatro Cienegas, Mexico, are home to ancient colonies of bacteria very similar to the ones that first arose on Earth over 3 billion years ago.

fields—you put all that together, and then you take the upper part of the ocean and mix that into the equation. That's about 3 percent of the biosphere, of the habitable part of the Earth. And the other 97% of that is cold and wet and dark and virtually unknown."

A significant portion of that 97% is made up of volcanic vents that harbor previously unknown life forms that exist in enormous quantities. According to Broad, "There's 46,000 miles [74,000 km] of them. That's like seven times around the Moon. It's the biggest feature on our planet, and we know virtually nothing about it, except that every once in a while when we go down . . . you find there are these lush ecosystems that thrive in total darkness."

When these volcanic sea floor areas and their life forms were discovered, people started to wonder if life on Earth might have started there. Broad says, "It's hot and it's constant. It's down there, cooking the whole time. It's like a test tube, a scientist's test tube. And the reasoning has evolved now . . . and evidence keeps building stronger and stronger that these places, these hot vents and these black chimneys—with all this hot—enormously hot water coming out—are the place where life began on this planet."

ASTROBIOLOGY'S MAIN QUESTIONS
To answer the questions posed in astrobiology, NASA has organized its research into a series of goals.

How Does Life Begin and Develop?
One of the key goals of astrobiology is to understand how life began on Earth; to explore how life evolves on the molecular, organism, and ecosystem levels; and to determine how the **terrestrial** biosphere has coevolved with Earth.

Does Life Exist Elsewhere in the Universe?

This is perhaps the biggest question for astrobiologists. Some goals for answering this question include determining what makes a planet habitable and figuring out how common habitable places are in the universe. Other goals include learning how to recognize the "**signature**" of life on other planets and to find out if there is, or ever was, life anywhere else in our solar system. Some places of special interest to astrobiologists are the planet Mars and also Jupiter's moon Europa.

What Is Life's Future on Earth and Beyond?

This question spurs the imagination. What will life on Earth be like in the year 10,000 or 50,000? NASA hopes to find out by studying how **ecosystems** respond to environmental change and understanding how terrestrial life responds to conditions in space or on other planets.

Chapter 2

History of Astrobiology

SCIENTISTS AND NONSCIENTISTS alike have speculated about the possibility of life on other planets for centuries. But since there was no means of exploring and testing their ideas, astrobiology did not become a science until fairly recently. Most of the first ideas of astrobiology were set down in the late 1950s: the notion that comets brought the first water and perhaps life to Earth from space, that nucleic acid was the first molecule that could replicate itself, and that Mars had life in some form.

Space technology was not developed enough during the 1950s to test and prove these theories: No space stations were available for deep-space observation; no space probes could orbit Mars; no rovers existed to travel over Mars's surface to send back samples. Astrobiology (which was called "**exobiology**" at the time) remained nothing more than an idea until space travel became a reality.

However, the delay in astrobiology research was not caused by the lack of technology alone. Many scientists in the 1950s thought exobiology was too much like science fiction. The only public discussion of life on other planets that was going on at the time happened in novels and fictional radio broadcasts. Scientists didn't want to risk not being taken seriously.

Without observable data, astrobiology lacked a firm foundation as a scientific discipline. On top of that, no one knew what to call it. Its name changed from *bioastronomy*, to *exobiology*, and finally, to *astrobiology*, as we know it today. In scientific literature, it was described as "life on other celestial bodies" or "**extraterrestrial** life."

Astronomers had noticed changing colors on the surface of Mars, which suggested to scientists that plant life might be growing there, perhaps being harvested by intelligent beings. There were also some ideas floating around the scientific community about comets. Scientists knew they were made of ice—frozen water—and so they were capable of carrying life around the universe. But aspiring astrobiologists needed technology to prove these theories, and that technology didn't exist.

NASA

When NASA was created in 1958, it was the first organization in the world to use rockets strictly for the purpose of space exploration and science. Up until then, rockets had been used only by the military. Beginning in the 1960s and throughout the 1970s, NASA launched many exploration

Figure 2.1 *Explorer 1*, seen here being launched on January 31, 1958, was the first American artificial satellite to orbit the Earth.

satellites into Earth's orbit, beginning with the *Explorer* series. These satellites had instruments that sent back information about the Earth's atmosphere. *Explorer 1* (Figure 2.1) made the major discovery that the Earth was surrounded by a radiation belt, which was later named the Van Allen Belt. The Van Allen Belt is a region of space where the Earth's magnetic field catches radiation, and it extends far into space. More discoveries were made by *Mariner 2*, which reached Venus on December 14, 1962, to scan its surface. *Mariner 2* measured Venus's surface temperature at 800°F (427°C). It also discovered what solar wind is made of.

Mariner 4 was launched on November 28, 1964, and reached Mars on July 14, 1965. It sent back 22 photographs of Mars's rough, reddish, rock-strewn surface. It also sent back data about Mars's atmosphere, which scientists discovered is made mainly of carbon dioxide. The *Lunar Orbiter* was launched in 1967. It orbited the moon and recorded scientific data about temperature, magnetic field, and surface chemistry. *Surveyor 1* made the first successful moon landing and returned to Earth with samples.

Throughout this time, technical advances were being made in **radio astronomy**. Computers were making it possible to process data from telescopes, and space travel got scientists thinking about the possibility of intelligent life in outer space. During the 1970s, after studying the moon, NASA sent *Mariner 6* to Mars, collecting data about its atmosphere. *Mariner 9* was sent into orbit around Mars in 1971. When it arrived there, it sent back information about a huge dust storm that lasted for a whole month. Once the storm cleared, *Mariner* was able to take photos of Mars's surface. These photos revealed giant inactive volcanoes, spectacular canyons, and dried riverbeds (Figure 2.2).

Figure 2.2 This spectacular image of the Olympus Mons volcano on Mars, the largest known volcano in our solar system, was taken by *Mariner 9*.

During the 1970s, Search for Extraterrestrial Intelligence (SETI) studies began. With radio telescopes aimed at the skies, scientists waited for signals from other planets. Scientists were now open to the idea that other technically advanced civilizations might be sending signals to let us know of their existence. This program lost its funding with NASA in the early 1990s, but SETI work was continued by the SETI Institute in California and by university programs around the country.

All of the data sent back to Earth from satellites and SETI spawned the modern field of astrobiology. The science of astrobiology could finally begin in earnest, because now there was information to research. The *Cassini-Huygens* orbiter was launched in 1996, a joint effort between NASA and the European Space Agency. *Cassini*'s goal, to reach Saturn and its moon Titan, has been accomplished.

When NASA established astrobiology research in 1996, it brought together scientists from many disciplines and formed a group called the NASA Specialized Center of Research and Training (NSCORT). This officially began the field of astrobiology at NASA.

Since then, many scientific discoveries have aided research in astrobiology. The Hubble Space Telescope has discovered planets around stars outside our **solar system**, which suggests that there may be other Earth-like worlds out there that could support life.

Here on Earth, scientists have also found life in extreme environments. Some of these extremes include high temperatures, high pressure, very salty or cold water, and high acidity. These extremes were once thought to prevent life. Now scientists have found that certain life forms not only survive but thrive within them.

Space exploration has provided scientists with evidence that liquid water has existed on Mars and on Jupiter's moon Europa. The fact that water was once present on an otherwise dry planet is a sign that life may have existed there at one time. Water must be available for life to happen.

New developments in technology such as **spectrometers**, ultraviolet and **infrared telescopes,** and radio telescopes have made it possible for astrobiologists to see what was once invisible to the eye or even with **optical telescopes**.

In 2005, the Mars rovers *Spirit* and *Opportunity* and their onboard instruments sent back data about Mars's atmosphere (Figure 2.3). Ground-based and space-based remote-sensing devices can detect the chemical signatures

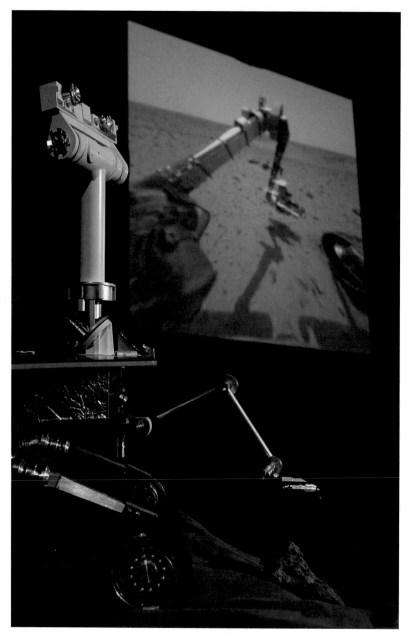

Figure 2.3 The Mars *Spirit* rover, modeled here at the Jet Propulsion Laboratory, is equipped to collect soil samples from the surface of Mars.

of atmospheres around other planets to determine whether life can exist on them.

Observations of comets and asteroids have supported the theory that life may have been brought to Earth by them. Comets and meteorites carry water and **microorganisms**.

Despite all the discoveries that have been made, the only life available to scientists for study is still that on Earth. NASA astrobiologist Chris Chyba says, "As long as Earth remains the only world on which life is known to exist, astrobiology has little choice but to build on our knowledge of terrestrial biology. We must always be aware that life in other places may have found other ways to make biology work, but exploring terrestrial life nevertheless gives us a starting point for many of the questions we ask."

Chapter 3

What Is an Astrobiologist?

AN ASTROBIOLOGIST IS A SCIENTIST who studies the living universe. An astrobiologist may have started his or her career as an engineer, mathematician, physicist, astronomer, or chemist. Studying the universe involves knowing its physical and biological qualities. It can't be studied with just one branch of science.

Astrobiology brings physical and biological sciences together to answer the basic questions of how life began and how life evolved here on Earth and elsewhere. It also tries to answer more specific questions such as: How do living systems emerge? How do habitable worlds form? How could terrestrial life survive and adapt beyond our home planet?

With the vast amount of information available thanks to advances in technology, scientists from many different fields now put their skills together to try to understand it all. This is what makes astrobiology **interdisciplinary**: Astrobiologists come from a variety of scientific disciplines.

WHAT ARE ASTROBIOLOGISTS LOOKING FOR?

Astrobiologists are looking for forms of life, both new and extinct. They are especially interested in life that can survive in extreme environments that may resemble conditions in outer space.

In 1977, scientists discovered a branch of life called **archaea** (Figure 3.1). Before this, there were only two known branches of organisms: **bacteria** (which do not have cells) and **eukaryotes** (animals, plants, and fungi, all of which have cells). Archaea are similar to both bacteria and

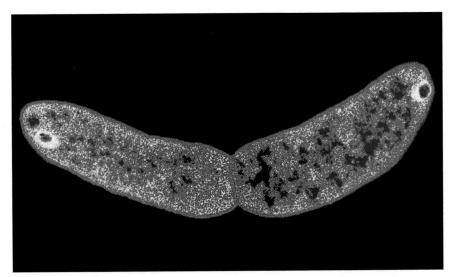

Figure 3.1 **This member of the archaea branch of living things is an extremophile that grows at very warm temperatures ranging from 30–37°C (86–98.6°F). It is seen here dividing.**

Archaea

The first archaea were discovered in 1977, and scientists thought they were a type of bacteria. After studying them more closely, however, they discovered that archaea were like bacteria, but were more closely related to eukaryotes (including humans). Scientists called them *archaebacteria* for a time, but to emphasize their distinct qualities, they now call them "archaea."

Extremophiles is another term scientists use when describing archaea. Extremophiles exist in different categories, according to their environments:

- thermophiles live at high temperatures
- hyperthermophiles live at extremely high temperatures (up to 121°C [250°F])
- psychrophiles live in cold temperatures
- halophiles prefer salty environments, like the Dead Sea
- acidophiles live in places with a very low pH (they like it at pH1, and will die if conditions reach pH7)
- alkophiles like a high pH.

Some 250 species of archaea have been discovered so far, and scientists have placed them into two groups: euryarchaeota and crenarchaeota.

Euryarchaeota live in swamps, sewage, cow stomachs, and termite guts. Other types of euryarchaeota like salty places (such as the Dead Sea) and acidic sulfur spots like the hot springs in Yellowstone Park.

Crenarchaeota like extremely hot environments. These are the hyperthermophiles. Other crenarchaeota like the acid, sulfur springs at Yellowstone. Some crenarchaeota make up the plankton in cool marine waters.

The importance of archaea lies in their ability to tell scientists about life in extreme environments both here on Earth and on other planets. But some people are also interested in their enzymes. Archaea enzymes might someday be used in detergents!

eukaryotes, but they live in extreme environments of intense heat, extreme cold, or with very little oxygen or light. Scientists call them extremophiles.

Because scientists have discovered that these organisms thrive here on Earth, they think that life-forms of this type (which require at least some water) might be found in the extreme environments of outer space, such as Mars.

EXTREMOPHILES IN YELLOWSTONE PARK

In 2003, University of Colorado (CU) researchers poked inside the rocks in the geyser regions of Yellowstone Park, where **geothermal heat** and high-acid water was once thought to be too extreme to support life of any kind (Figure 3.2). This area, known as the Norris Geyser Basin, is very acidic and is also the hottest geyser area in Yellowstone Park and maybe the world. Yet tiny living creatures— **microbials**—find this environment a good place to live and grow. "The pores in the rocks where these creatures live have a **pH** value of one, which dissolves nails," one CU researcher said in a 2005 University of Colorado news release.

Doctoral student Jeff Walker helped write a research paper on the topic. He discovered this community of microbes when he broke apart a chunk of sandstone one day at the Norris Geyser Basin. "I immediately noticed a distinctive green band just beneath the surface," he said. "It was one of those 'eureka' moments."

Research at the Norris Geyser Basin has shown that **fossils** left by these organisms have distinctive imprints, and could serve as signatures for ancient life in geothermal environments on Earth and on other planets in our solar system.

Astrobiologist Lynn Rothschild has also done research at Yellowstone Park. She describes the acid hot springs of

Figure 3.2 The hot springs at Yellowstone Park are some of the most forbidding places on Earth, yet scientists have found organisms that actually thrive in the hot, acidic conditions.

Yellowstone as so unfriendly to humans that they "would make a witch's cauldron seem benign. And yet they teem with life." The life to which she is referring includes organisms known as **thermoacidophiles**—microbes that love hot, acidic places.

In Octopus Springs in Yellowstone Park, Rothschild performed research on a **microbial mat**. These mats show how microbial life adjusts to the conditions of the environment. The organisms on the top layers of mats use sunlight for **photosynthesis**, but in thicker areas of the mat, the light of the sun is blocked. Cells in these areas are replaced by microbes that can survive in dimmer light. Eventually, at the bottom, organisms adapt to live without any light at all.

EXTREMOPHILES DEEP IN THE EARTH

When she isn't checking out geysers, Lynn Rothschild is examining slime. Actually, it is SLiME, which in biology stands for *Subsurface Lithoautotrophic Microbial Ecosystem*. SLiME organisms live deep in the Earth and are extremophiles. Since they live far below the Earth's surface—in some cases, 2 miles (3.2 km) down—they don't receive energy from the sun for growth. They get their nutrients from rocks in a process called **chemosynthesis**.

SLiME life forms have been found in solid lava more than a mile underground. The farther down these microbes live within the Earth, the more extreme the environment is. Scientists estimate that some SLiME communities may be several million years old. If SLiME organisms can't find enough nutrients to live on, they become smaller, shrinking down to as little as one-thousandth of their normal size. These reduced-sized microbes are still alive, but they are starving. With so few nutrients to support them, they reproduce very infrequently, maybe once in 100 years.

Microbial Mats

Microbial mats are an active **biogeochemical** zone a few millimeters thick. They're found near geysers and marshes by lakes and oceans. They produce their own energy.

Some of the earliest forms of life on Earth were bacteria that stuck together in layers to form microbial mats. Scientists think microbial mats nearly covered the planet at one time. Microbial communities (in places like the Yellowstone geysers) are where astrobiologists are currently doing most of their research. If they can understand these life forms and how they evolve, it will help them recognize and identify microscopic life on other planets.

EXTREMOPHILES IN THE SEA

Astrobiologists want to learn more about how life began. To do this, they often study organisms in the sea, because they believe that these organisms are going through a process similar to what was happening when life began on Earth billions of years ago.

Far below the surface of the ocean there are **fissures**, where hot lava from deep within the Earth has broken through the ocean floor, releasing extremely hot columns of mineral-filled water. These "black smokers," or hydrothermal vents, are where extremophiles called thermophiles are found. They love the heat.

Thermophiles, like SLiME organisms and other types of extremophiles, do not get their energy from the sun, but instead from chemical reactions. Thermophilic bacteria and other life-forms have adapted to this environment and live off the dissolved metals and deposits that come from eruptions.

EXTREMOPHILES IN THE MOUNTAINS

High above sea level at 19,680 feet (5,999 meters), on the border of Chile and Bolivia, South America, there's a frigid lake in the crater of Licancabur, a volcanic mountain peak. This lake doesn't have a name, but it is famous for being the highest lake in the world. Here, the environment is extreme. The lake is usually frozen, and the oxygen level is very low, about half of that found at sea level. The high altitude exposes the lake to high levels of **ultraviolet radiation** from the sun.

Extremophiles have made themselves at home here. Scientists study these microbes because they want to learn about how life might survive on Mars, which also has dry lakes, low oxygen levels, and high ultraviolet radiation.

Astrobiologists Nathalie Cabrol and her husband, Edmond Grin, travel to these extreme environments, because they say they want to learn more about Mars: "[L]ife is thriving in conditions that relate very much to Mars three and a half billion years ago."

NEW PLANETS AND STARS

Astrobiologists are also studying the newly discovered planets outside of our solar system, which currently number more than 100. They're wondering if it is possible for Earth-like orbiting planets to harbor life. A team of astronomers from the Geneva Observatory in Switzerland have discovered a planet with an orbit path very similar to Earth's. The orbit of a planet says a lot about whether it could support life. For example, the mostly circular orbit of the Earth keeps it in a stable habitable zone (scientists sometimes refer to this as the "Goldilocks zone"—not too hot, not too cold, but just right) where the warmth of the sun allows life to thrive.

If a planet's orbit is too **elliptical**, distances can be too extreme at some points—either too close or too far away from the star that gives the planet warmth. Earth-like orbits are also of interest to astrobiologists because the stable temperatures would allow any water on the planet to remain there and not evaporate. Finding planets with liquid water is one of the key goals of astrobiology.

METEORITES AND COMETS

Scientists have found evidence of bacteria-shaped "fossils" in a meteorite from Mars, named ALH84001. There is evidence of magnetite crystals on this meteorite, similar to magnetite on Earth. Laboratory experiments are currently

Asteroids, Comets, Meteors, and Meteorites

After the sun, moon, and planets of our solar system were formed, a lot of building pieces were still left over. These pieces are comets and asteroids. Although they usually hang out in the Asteroid Belt between the orbits of Mars and Jupiter, they sometimes head straight for the sun. Asteroids are mostly made of rock, and comets contain a lot of ice. They warm up on their way to the sun, which vaporizes their icy bodies and creates a long streaming tail of dust and gas.

Asteroids are chunks of stone or metal or both. They can be the size of boulders or even bigger than mountains. When they collide, they shatter, creating fragments that fall into orbits around the Earth. Smaller fragments are called meteoroids, which usually vaporize into meteors. When a meteor reaches Earth, it is called a meteorite (like the one the students below are touching). These particles often land on Earth. About 50 tons of meteorite material hit the planet every day!

under way to determine the origin of these crystals and to find out whether they are an indicator of life on Mars.

Scientists have determined that comets contain materials from the time when the solar system was first formed. Astrobiologists examine comets at varying distances from the sun to study their structure and the composition of the comets' dust and ice. They compare this information with similar dust grains and ice particles found in other realms, including Earth.

Chapter 4

Tools of the Astrobiologist

ASTROBIOLOGY DEPENDS A GREAT DEAL on the tools of technology. When space technology became advanced enough to send information back to Earth about planets and stars, astrobiology went from being a theory to a real science with something to study, explore, and research. The data that astrobiologists use are gathered by space probes, satellites, rovers, ground and space-based telescopes, observatories, spectrometers, interferometers, and photographic equipment. And this is just a short list of the tools of astrobiology!

Astronomical observations can be divided into two basic types: **direct imaging** and **spectroscopy**. The astrobiologist uses both.

SPACE PROBES

The information scientists received from the earliest *Explorer* space missions in the 1960s and 1970s provided data about the moon, solar wind, and the Van Allen Belt, and was a foundation for astrobiology. These early missions taught scientists something about how Earth's near-space environment worked.

Deep-space observation satellites, such as *Mariner 4*, provided information about Mars's surface and atmosphere. *Mariner 9*'s visit to Mars sent back information about the possibility that water once existed on Mars, suggested by photos of dried-up riverbeds. These discoveries spurred the interest of NASA and eventually led the space agency to build the *Viking* aircraft.

NASA focused on Mars for deep-space exploration in 1975, launching *Viking 1* and *Viking 2*, each of which had an orbiter and a lander. The orbiters both monitored Mars's weather and mapped out its surface. The landers scooped up soil samples, which were analyzed right there, on Mars, within the landers' onboard laboratories.

In the 1990s, *Cassini-Huygens* and the Mars Global Surveyor were launched. *Cassini-Huygens* was sent to Saturn, and carried the Huygens probe on board, to be released into the moon Titan's atmosphere.

"Titan is a time vault," writes one scientist. "Its unique environment and thick atmosphere may resemble that of Earth some several billion years ago, before life as we know it began pumping oxygen into our atmosphere." Titan is of great interest to scientists because it is the only moon known to have clouds and a mysterious, planet-like atmosphere. Titan's atmosphere is made up mainly of nitrogen, which appears as an opaque orange haze, hiding its surface—and its secrets—from view. The *Cassini-Huygens* mission will try to reveal those secrets during its four-year tour of Saturn, its moons, and its **magnetosphere**. Once they get a better understanding of Titan's surface, atmosphere, and chemical composition, scientists hope to learn more about what early Earth might have been like billions of years ago.

The Mars Global Surveyor (MGS) began orbiting Mars on September 12, 1997, and its camera has now examined nearly 4.5% of Mars's surface. The MGS also has a wide-angle camera that observes the entire planet every day (Figure 4.1).

Figure 4.1 The antenna on this Mars lander is what transmits signals to the Mars Global Surveyor as it orbits. The MGS then sends the information it receives to control centers here on Earth.

The MGS has produced images for a longer period of time than any other spacecraft ever sent to Mars, surpassing *Viking Lander 1* in 2004. It has "returned more images than all past Mars missions combined," said Tom Thorpe, project manager for MGS at NASA's Jet Propulsion Laboratory in Pasadena, California.

HUBBLE SPACE TELESCOPE

In 1962, the National Academy of Sciences recommended that a large space telescope be built. In 1977, Congress voted to fund the project and construction of the Hubble Space Telescope began (Figure 4.2).

In June 1994, NASA released Hubble Orion **Nebula** images that confirmed the births of planets around newborn stars. In November 1995, NASA released Eagle Nebula images that also showed where stars are born (Figure 4.3). In January 1996, NASA released "Deep Field" images in which Hubble peered back in time more than 10 billion years. The Hubble Space Telescope so far has revealed at least 1,500 galaxies at various stages of development.

On the Hubble Website, a NASA scientist described the sight of starbirth within the Eagle Nebula:

A torrent of ultraviolet light from a band of massive, hot, young stars [off the top of the image in Figure 4.3] is eroding the pillar. The starlight also is . . . illuminating the tower's rough surface. Ghostly streamers of gas can be seen boiling off this surface, creating the haze around the structure and highlighting its three-dimensional shape. The column is silhouetted against the background glow of more distant gas. In this celestial case, thick clouds of hydrogen gas and dust have survived longer than their surroundings in the face of a blast of ultraviolet light from the hot, young stars.

Figure 4.2 The Hubble Space Telescope has become one of the most useful tools for astrobiology and for space science in general.

The Hubble Space Telescope has been significant to astrobiology. It has also been instrumental in the recent discovery of planets outside our solar system and new galaxies in deep space. After 15 years and more than 700,000 observations, Hubble has been the source of seemingly endless scientific information. It has allowed scientists to observe planets in our solar system very closely, and to learn more about how these planets change over time.

SPECTROMETERS

Spectrometers are instruments used by astronomers and astrobiologists to identify materials in space. By heating up

Figure 4.3 **The image of starbirths taking place in the Eagle Nebula was taken by the Hubble Space Telescope in November 2004.**

a sample of material taken from the surface of the planet, the spectrometer makes the sample give off light at a particular **frequency**. Different materials radiate energy at frequencies that are unique to their chemical makeup, somewhat like a fingerprint. Spectrometers broadcast these frequencies. Space labs onboard rovers perform these chemistry experiments and send the results back to Earth.

Spectrometers go through a process of spreading light out into **wavelengths**, creating a **spectrum**. Within this spectrum are **emission** and absorption lines, the fingerprints of atoms and molecules. These show scientists what a material is made of. Each atom has a unique spacing of orbits and can emit or absorb only certain energies or wavelengths. This is why the location and spacing of lines is unique for each atom.

Astronomers can learn a great deal about an object in space by studying its spectrum, such as what it is made of, its temperature and density, and its motion (both its rotation and how fast it is moving toward or away from us).

KECK INTERFEROMETER

The Keck Interferometer is a system that links two 10-meter (33-foot) optical telescopes, creating the world's most powerful optical telescope system. The Keck Interferometer is used to search for planets around nearby stars. It combines light from the two telescopes to measure the emissions from dust orbiting nearby stars.

INFRARED TELESCOPES
Spitzer Space Telescope

The Spitzer Space Telescope uses infrared spectrometers to collect information about the surfaces of planets. Telescope

Spitzer Space Telescope

The Spitzer Space Telescope fills in an important gap in wavelength coverage not available from the ground—the thermal infrared.

The Spitzer Space Telescope was launched on August 25, 2003. Spitzer obtains images by detecting the infrared energy, or heat, given off by objects in space between wavelengths of 3 and 180 microns (1 micron is one-millionth of a meter).

Spitzer is the largest infrared telescope ever launched into space. It has highly sensitive instruments that allow scientists to peer into regions of space that are hidden from optical telescopes. Many areas of space are filled with vast, dense clouds of gas and dust that block our view. Infrared light can penetrate these clouds, letting scientists look into regions of star formation, the centers of galaxies, and newly forming planetary systems. Spitzer took this image (below) of the remnant of a supernova.

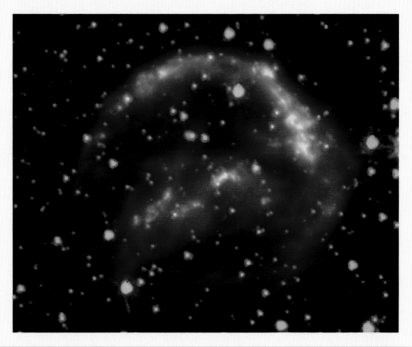

sensors record the information with infrared images, which are beyond visible wavelengths. Colors are then assigned to the different wavelengths, and these variations in color make the image visible. The colors indicate climate changes and surface biology.

The NASA Infrared Telescope Facility at the Institute for Astronomy at the University of Hawaii has recently been able to determine that there is methane gas on Mars through a close reading of infrared images. Methane gas is a by-product of microbial life forms and is associated with the presence of life.

Chapter 5

What Do Astrobiologists Do?

WHERE DO ASTROBIOLOGISTS WORK?

IF YOU'RE INTERESTED in astrobiology, you're in luck, because this new field of science is really booming. Many colleges and universities around the country now have programs in astrobiology, so training should be easy to find. And once you graduate, you might find a job with one of the lead teams of NASA's Astrobiology Institute, or you can check out the SETI Institute's astrobiology projects. PlanetQuest and the other branches of NASA—including the Ames Research Center, the Jet Propulsion Laboratory, and NASA headquarters in Washington, D.C.—all hire astrobiologists. University astrobiology departments and observatories also have research and management positions. Astrobiologists might work in a laboratory or in the community, teaching people about life in the universe at science museums and planetariums.

NASA ASTROBIOLOGY INSTITUTE

The NASA Astrobiology Institute (NAI) is a **virtual** institute. A "virtual" institute is a group of scientists who conduct research at their own college, university, or government sites. They work together, but not in the same place, or even the same state. They use technology to communicate their ideas and research findings.

Part of the job of NAI teams is to train a new generation of researchers in the field of astrobiology and to reach out to teachers, students, and the general public.

The NAI sponsors focus groups for specific subject areas. These focus groups are open to non-NAI members. The NAI has partnerships with international astrobiology organizations to help cultivate cooperation and interaction in astrobiological research around the world.

A NASA Astrobiology Postdoctoral Fellowship Program, administered by the NAI in association with the National Research Council, selects a group of the best young astrobiology researchers each year and supports them directly.

Astrobiologists frequently work in teams. At NASA, astrobiology teams work in focus groups. Within these focus groups, astrobiologists do different jobs, depending on their assignments and their scientific training.

NAI Lead Teams

Ames Research Center, Mountain View, CA
http://www.arc.nasa.gov

Carnegie Institution of Washington, Washington, D.C.
http://astrobiology.ciw.edu

Indiana University, Bloomington, IN
http://www.indiana.edu/~deeplife

Marine Biological Laboratory, Woods Hole, MA
http://astrobiology.mbl.edu

Michigan State University, East Lansing, MI
http://astrobiology.msu.edu/index.html

NASA Goddard Space Flight Center, Greenbelt, MD
http://astrobiology.gsfc.nasa.gov

Pennsylvania State University, University Park, PA
http://psarc.geosc.psu.edu

SETI Institute, Mountain View, CA
http://www.seti.org/seti_nai

University of Arizona, Tucson, AZ
http://www.arizona.edu

University of California at Berkeley, Berkeley, CA
http://cips.berkeley.edu/biomars/NAI_1.html

University of California at Los Angeles, Los Angeles, CA
http://www.astrobiology.ucla.edu

University of Colorado at Boulder, Boulder, CO
http://argyre.colorado.edu/life

University of Hawaii (Manoa), Honolulu, HI
http://www.ifa.hawaii.edu/UHNAI

University of Rhode Island, Kingston, RI
http://www.gso.uri.edu/astrobiology

University of Washington, Seattle, WA
http://depts.washington.edu/astrobio

Virtual Planetary Laboratory, Pasadena, CA
http://vpl.ipac.caltech.edu

ASTROBIOLOGY FOCUS GROUPS AT NASA
Mars Focus Group

The Mars Focus Group contemplates past or present life on Mars and **prebiotic** (pre-life) chemistry. This focus group makes recommendations and gives advice about upcoming astrobiology missions.

Mission to Early Earth Focus Group

The search for life beyond Earth requires an understanding of the conditions where life originates and evolves. This understanding is based on Earth, the only planet where life is known to exist. The Mission to Early Earth Focus Group believes that the study of life and the environment of the early Earth is a critical part of developing plans for future astrobiology space missions.

EvoGenomics

The purpose of the Evolutionary Genomics (EvoGenomics) Focus Group is to research **evolutionary genomics** as it relates to astrobiology. Analyzing the evolution of the **genomes** of organisms has advanced our understanding of how life began, how it adapted to diverse environments, and how it became more complex on this planet. NASA expects that these studies will lead to a better understanding of life elsewhere in the universe. The EvoGenomics Focus Group is a unique collaboration of astrobiologists with expertise in molecular evolutionary analysis, organic chemistry and biochemistry, Earth history, and paleontology. According to the NASA Website: "Our unifying goal is to compare the early evolutionary history of life, as revealed through analyses of genomic sequence data, with changes in Earth's environment through time, providing the basis to identify biomarkers [signatures] for habitable planets."

Europa

The Europa Focus Group promotes the expansion of scientific studies and joint investigations of one of Jupiter's moons, Europa, to help scientists understand its potential for supporting life (Figure 5.1). Scientific study of Europa is interdisciplinary. It uses spacecraft data analysis, along with laboratory simulations and other techniques, to learn more about the history and present environment in and around Europa.

Europa

Europa has been identified by NASA and the National Academy of Sciences as a high priority for exploration, because it appears to be one of the few bodies in our solar system with conditions that are favorable for the development of life.

Europa is a rocky object, slightly smaller than our moon. It has an outer shell made up of water, the surface of which is frozen. Scientific evidence suggests that liquid water existed below Europa's icy crust in the recent geologic past and that liquid water might still be present today.

Friction within Europa is created by tides and by interaction with Jupiter and neighboring moons. This friction might generate enough heat to create volcanoes at the base of the water layer. These factors and remote-sensing observations of salts on the surface and the presence of organic compounds (brought by comets) lead scientists to believe that Europa may have all the ingredients needed for life: an energy source, water (possibly liquid), and organic chemistry.

Figure 5.1 Jupiter is seen here with its four moons. Europa, which astrobiologists believe may be able to support life, is the small sphere at top right.

Impacts

The Impacts Focus Group studies how the impacts of asteroids, comets, and other materials have influenced the origin, evolution, and extinction of life on Earth. A major goal of this group is to gather samples from a wide variety of places (both on Earth and from outer space). As NASA's focus groups Website explains, impacts have affected the landscapes of "every rocky or icy planet in the solar system. If life ever existed on astrobiological points of interest such as Mars or Europa, then impacts have likely affected the biota on these bodies as they have done on Earth."

Astrobiotech

The Astrobiotech Focus Group's goal is to identify the science and technology needs for the unique research of astrobiology. NASA's exploration of the solar system involves the search for signatures of extraterrestrial life. To accomplish this, advances in *in situ* (which means "in a natural location") science are needed, including landing instruments to take samples and analyze their physical and chemical properties.

Instrument technologies for in situ planetary exploration could be used to study extreme environments on Earth as well, such as deep-sea vents, high-altitude environments, and locations in the Arctic Sea and Antarctica.

To address the challenges of exploration, the Astrobiotech Focus Group's goal is both to identify gaps in the technology of Mars missions and to transfer the methods of Earth-based experiments to other planets.

WORKDAY OF AN ASTROBIOLOGIST

Astrobiologists may spend the day in a laboratory, creating simulated environments to test the possibilities of life growing within them. They may work with lasers, telescopes, **mass spectrometers**, and computers. Astrobiologists collect field samples in extreme environments and use their training in biology, physics, and chemistry to explain and understand the nature of what they find.

Astrobiology experiments, such as how life responds to conditions in space, are conducted at the International Space Station and on other space missions. These experiments may take years of planning.

Astrobiologists work on research. They build on discoveries made by other scientists and apply these theories and principles to their own scientific study.

Chapter 6

Becoming an Astrobiologist

TRAINING FOR ASTROBIOLOGISTS

If you're interested in a career in astrobiology, you should take as many math and science classes as you can. This will give you a strong background for courses in biology, physics, and advanced mathematics in college. These are the foundations of astrobiology training. Scientists use the principles of math, physics, and chemistry to understand life in the universe.

INTERNSHIPS

Getting an internship at an organization can lead to a job. But even if you don't want to start work right away, internships are a great way to learn the ropes in the real world. NASA sponsors many internship programs for people who might like to become astrobiologists (see the box on page 50).

Research projects for astrobiology interns vary. Students may work on developing instruments designed for analyzing organic materials in situ, synthesizing and analyzing cosmic ice or smoke in a laboratory, searching for and ana-

College and University Programs in Astrobiology

The following universities offer four-year degrees in astrobiology.

University of Arizona
http://www.arizona.edu

University of California at Berkeley
http://cips.berkeley.edu/biomars/NAI_1.html

University of California at Los Angeles
http://www.astrobiology.ucla.edu

University of Colorado at Boulder
http://argyre.colorado.edu/life

University of Hawaii (Manoa)
http://www.ifa.hawaii.edu/UHNAI

Indiana University
http://www.indiana.edu/~deeplife

Michigan State University
http://astrobiology.msu.edu/index.html

Pennsylvania State University
http://psarc.geosc.psu.edu

University of Rhode Island
http://www.gso.uri.edu/astrobiology

University of Washington
http://depts.washington.edu/astrobio

lyzing atmospheres, and analyzing infrared spectra from planetary bodies in search of organic molecules.

WHAT KIND OF PERSON BECOMES AN ASTROBIOLOGIST?

Astrobiologists are inquisitive and adventurous. Many astrobiologists are daring, and they take considerable risks to pursue their desire to learn about the origins of life. They travel to extreme environments deep in the ocean in tiny submersible vessels and climb to frozen mountain areas to look at small, even microscopic, organisms.

Astrobiologists are good at problem-solving and analyzing information. They have broad imaginations, an important quality in a person who thinks about finding life in outer space. A necessary trait in an astrobiologist is initiative, to help transform ideas from the imagination into reality.

Astrobiology Internships

Lunar and Planetary Institute

NASA Academy

NASA Astrobiology Summer Undergraduate Internships

NASA Langley Aerospace Research
Summer Scholars Program

NASA Planetary Geology and Geophysics
Undergraduate Research Program

NASA Robotics Internship Program

NASA Undergraduate Student Research Program

Space Grant Internship Program at Jet Propulsion
Laboratory

Astrobiology is a relatively new field, and lots of people want to know more about it. Having good communication skills, both in writing and orally, are important for an astrobiologist who wants to further the understanding of the field and to educate the public.

Most astrobiologists have a Ph.D. in another discipline, which they apply to the study of life in the universe.

PROFILE OF AN ASTROBIOLOGIST

Pamela Conrad Gales took the long way to find her true career passion, astrobiology (Figure 6.1). You might say it is her second career. She studied music in college and got a master's degree in music composition because she wanted to be a composer. After 18 years in the broadcasting business, where she worked as a producer and engineer, she decided to change directions. She said, "One day, I just decided to be a scientist and so began the tortuous process of getting the proper academic credentials."

Her subdiscipline in astrobiology is geobiology. When she decided to pursue this field, she says, "Geobiology didn't really exist yet, but George Washington University, in Washington, D.C., did have a program in geobiology [studies]," but it focused mainly on paleontology and anthropology. Since she wanted to study geobiology more than these other subjects, she decided to get "a solid background in geology and just become a geobiologist" after getting her Ph.D.

Around the time she was finishing her degree, NASA began its astrobiology program. "A well-known microbiologist named Ken Nealson was just starting up a research group at JPL [the Jet Propulsion Laboratory] and he hired me to be a scientist in his group. The day I arrived he said, 'I need you to give a talk in my place about what astrobiology

Figure 6.1 Astrobiologist Pamela Conrad Gales is seen here taking a break while exploring Antarctica.

is.'" She was to give the talk to the entire JPL community. "I said 'OK', but it took a lot of scrambling to figure out what astrobiology is, so I could talk about it!"

As a NASA astrobiologist, Gales spends a lot of time in the field working on science projects, "mostly in deserts, especially cold ones in the arctic, and even in Antarctica. This is what I like best about my job—I get to be an explorer in some cool places and learn how they might be relevant to places in the solar system. . . . We discover something new each time we go out."

Gales says, "Everyone can be an explorer. It takes no special skill, just curiosity and a love of discovery. If you learn the scientific method, then you can be both an explorer and a detective."

List of Current Job Titles and Occupations on File at NASA

A background in physics, biology, mechanical and electrical engineering, math, software engineering, or astrobiology qualifies applicants for many of these positions.

Advanced Projects Design Team Leader

Aeroacoustics Engineer

Aerodynamics Engineer

Aeromechanics Engineer

Aeronautical Engineer

Aeronautics and Aerospace Technologist

Aerospace Engineer

Aerospace Engineer Technician

Aerospace Optical Engineer

Aerospace Research Engineer

Aerospace Systems Safety Research Assistant

Aerospace Technologist

Airborne Telescope Operator

Analytical Chemist

Applied Meteorology

Assistant Astronomer

Assistant Branch Chief

Assistant Science Coordinator

Assistant Superintendent

Associate Producer

Associate Staff Scientist

Astrobiologist

Astronaut

Astronomer

Astronomy Educator

Astrophysicist

Atmospheric Physicist

Atmospheric Structure Investigator

Biocomputation Center Deputy Director

Bioengineer

Biological Engineer

Biologist

Biomedical Engineer

Biomedical Technician

Branch Chief

Center Controller

Chemical Engineer

Chemist

Chief of Biological & Chemical Analysis Laboratories

Chief of Guidance and Propulsion Systems

Chief of Life Sciences Division

Chief Project Engineer

Chief Toxicologist

Cinematographer

Civil Engineer

Cognizant Engineer

College Intern

Commander

Computational Fluid Dynamicist

Computational Fluid Dynamics Engineer

Computer Engineer

Computer Programmer

Computer Scientist

Computer Technician

Conceptual Aircraft Designer

Congressional Staff Member

Crew Chief

Crew Coordinator

Crew Training

Curriculum Specialist

Data Communications Engineer

Data Management Team

Data Systems Specialist

Deep Space Tracking Network Operations Project Engineer (NOPE)

Deputy Chief of Propulsion and Fluid Systems Branch

Deputy Chief, Systems Division Mission Operations Directorate

Deputy Director of Aeronautics

Deputy Director of Operations, Research and Development Services

Deputy Manager of Payloads Office

Deputy Navigation Team Chief

Deputy Uplink Systems Engineer

Design Engineer

Design Lead

Development Group Leader

Director of Development, Space Center Houston

Director of Flight Operations

Director of Public Relations

Director, Astrobiology and Space Research

Director, California Air & Space Center Teacher Institute

Director, Counseling and Psychological Services Center

Director, NASA Life Sciences Division

Dive Specialist, Neutral Buoyancy Lab (NBL)

Division Chief

Education Specialist

Electrical Supervisor

Engineer

Engineering Analyst

Engineering Assistant, NASA's SHARP Program

Engineering Manager

Engineering Technician

Engineering Test Pilot

Environment Control and Life Support Systems Engineer

Environmental Physiologist

Environmental Protection Specialist

Environmentalist

Exercise Physiologist

Exobiologist

Experiment Integration Engineer

Experiment Processing Engineer

Experiment Support Scientist/ Microbiology Coordinator

Experiment Systems Manager

Geophysicist

Hardware Engineer

Hazardous Robotics Specialist

High-Energy Astrophysicist

HST Astronomer

Journalist

K–12 Outreach Support Personnel

Knowledge Engineer

Laboratory Manager

Lander Camera Support Personnel

Launch Site Support Office Personnel

Launch to Activation Procedures Lead

Lead Altimetry Analyst

Lead Mechanical Engineer

Lead Mechanical Technician

Lead Ops planner, Mission Operations Directorate

Lead Robotics and Avionics Engineer

Lead Schedule Integration Engineer

Lead Shuttle Systems Inspector

Lead, Space Station Power Resource Management Team

Leader of the Test Engineering Group

Life Science Space Experiment Ground Lab Logistics Coordinator

Life Science Specialist

Life Sciences Division Deputy Chief

Life Sciences Education Programs Coordinator

Life Sciences Outreach Office Personnel

Life Sciences Program Manager

Local Controller

Logistics Operations Manager

Magnetometer (MAG) Science Coordinator

Manager of Mars Exploration Program

Manager, Galileo Administrative Office

Manager, Mars Sample Return Lander

Manager, Space Shuttle Office

Mars Atmosphere Interdisciplinary Scientist

Mars Exploration Program Architect

MARS Outreach Program Project Coordinator

Materials Engineer

Materials Scientist

Mathematical Researcher

Mechanical Engineer

Mechanical Instrumentation Tech

Mechanical Technician, Shuttle Systems

Microbial Ecologist

Mission Commander

Multimedia Specialist

National Research Council Post-Doctoral Fellow

National Science Foundation Representative, Palmer Station

Navigator

Network Engineer

Neurobiologist

Neuroscience Researcher

New Space Transportation Developer

NTSB/FAA Investigator

Numerical Software Engineer

Nutritionist

Observational Infrared Astronomer

Observing Assistant

Oceanographer

Operations Lead

Optical Engineer

Orbital Debris Scientist

Orbiter Operations Group Lead

Orbiter Processor

Orbiter Test Conductor

Ornithologist

Outreach Program Manager for Life Sciences

Outreach Specialist

Photopolarimeter Radiometer Instrument Engineer

Photopolarimeter Radiometer Science Coordinator

Physicist

Physics Research Associate

Physiologist

Pictures/Remote Sensing Specialist

Pilot

Planetary Geologist

Planetary Scientist

Plasma Wave Assistant Science Coordinator

Power, Heating, Articulation, Lighting and Controls Officer

Principal Investigator

Principal Scientist

Probe Deputy Manager

Professor

Program Manager

Program Planning Specialist

Program Scientist

Project Engineer

Project Manager

Project Scientist

Propulsion Engineer

Psychophysiologist

Radio Astronomer

Real-Time Operations Lead

Research Instrument Maker

Research Assistant

Research Associate

Research Astrophysicist

Research Engineer

Research Nutritionist

Research Physicist

Research Pilot

Research Scientist

Science Instruments Specialist

Science Planning Coordinator

Sciences Requirements Manager

Scientific Director

Scientist

Sequence Integration Engineer

Shuttle Structures and Transporters Engineer

Shuttle Test Director

Simulation Supervisor

Software Development Lead

Software Engineer

Solar Physicist

Solar Scientist

Space Farming Engineer

Space Flight Technician

Space Flight Training Specialist

Space Physicist

Space Scientist

Space Shuttle Crew

Space Shuttle Remote Manipulator System Training Instructor

Space Station Robotics Instructor

Space Suit Project Engineer

Spacecraft Design Engineer

Spacecraft Systems Engineer

System Safety, Reliability & Quality Assurance Lead

Systems Engineer

Systems Management

Systems Verification

Teacher

Team Manager, ISS Mission Evaluation

Technical Advisor, Earth: Final Conflict Television Series

Technical Integration Engineer with Space Station

Technical Leader for Space Station Outreach Group

Technical Writer

Technician

Technology Transfer Specialist

Test Engineer

Thermal Protection System/Shuttle Upgrades Specialist

Trajectory and Aerobraking Design Analyst

Trajectory Optimization Engineer

Archaea: A branch of organisms that are neither bacteria nor eukaryotes, but similar to both.

Astrobiologists: People who study life in space, on planets, stars, moons, and comets.

Bacteria: Single-celled organisms seen only with a microscope.

Biogeochemical: Short for "biological, geological, and chemical;" refers to the recycling chemistry between plants, animals, and the sediments of the Earth.

Biosphere: The portion of the Earth and its atmosphere that can support life.

Black smokers: *See* **Hydrothermal vents**.

Chemosynthesis: The process in which some bacteria use chemicals (like hydrogen sulfide) to obtain the energy they need for life.

Direct imaging: The use of a telescope as a camera to create images of celestial objects.

Ecosystems: The natural systems in which energy and nutrients cycle among plants, animals, and their environment.

Elliptical: Oval-shaped.

Emission: To discharge a substance.

Eukaryotes: A branch of organisms that includes plants, animals, and fungi.

Evolutionary genomics: The study of how the genetic codes of organisms change over time.

Exobiology: Another word for "astrobiology."

Extraterrestrial: Something that originates, is located, or occurs outside Earth or its atmosphere.

Extremophiles: Organisms that are able to survive in harsh environments, such as those that lack oxygen or light, or are extremely cold or hot.

Fissures: Cracks.

Fossils: Traces left in rock or other material of organisms that were once alive.

Frequency: How often a wavelength completes a cycle over time.

Genomes: The genetic makeup of an organism.

Geothermal heat: Warmth arising from the inside of the Earth.

Habitable: To be suitable for life.

Hydrothermal vents: Openings in a planet's surface where geothermally heated (heated underground) water emerges; also called "black smokers."

Infrared telescope: A telescope designed to detect radiation in the infrared range of the electromagnetic spectrum.

Interdisciplinary: Using contributions from several different fields.

Magnetosphere: The magnetic field around a planet, located above the planet's uppermost layer of atmosphere.

Mass spectrometers: Instruments that identify the chemical makeup of a substance.

Microbes: *See* **Microorganisms**.

Microbial: Of or relating to microbes (microorganisms).

Microbial mat: Mats that show how microbial life adjusts to the conditions of the enivronment.

Microorganisms: Also called "microbes;" organisms that can be seen only with a microscope.

Nebula: Interstellar cloud of dust and gases.

Optical telescopes: Traditional telescopes that use a series of lenses and mirrors to obtain images of bodies in space.

pH: Potential (of) hydrogen; a measure of the activity of hydrogen ions (H+) in a solution, which indicates acidity or alkalinity.

Photosynthesis: The process in plants and certain organisms which produces carbohydrates from CO_2 and water using light.

Prebiotic: Referring to the conditions before life began to form.

Radio astronomy: The study of the universe with radio telescopes.

Satellites: A moon or manmade object revolving around a planet.

Signature: Sign or indication of life, present or past.

Solar system: A group of planets orbiting around a star.

Spectrometers: Instruments for measuring the intensity of radiation as a function of wavelength.

Spectroscopy: The examination of spectra.

Spectrum: Range of colors or frequencies.

Submersible: Vessel able to explore underwater.

Terrestrial: Of the Earth.

Thermoacidophiles: Organisms that thrive in hot, acidic environments.

Ultraviolet radiation: Energy emitted in the form of light from the ultraviolet segment of the electromagnetic spectrum.

Virtual: Something that exists in effect, but not in reality.

Wavelengths: The distance a light wave travels over time.

Corbella, Luciano. *Is Anybody Out There?* New York: DK Publishing, 1998.

Jefferis, David. *Alien Lifesearch: Quest for Extraterrestrial Organisms.* New York: Crabtree Publishing, 1999.

Nardo, Don. *Extraterrestrial Life.* Farmington Hills, MI: Lucent Books, 2004.

Sherman, Joseph. *Deep Space Observation Satellites.* New York: Rosen Publishing Group, 2003.

Skurzynski, Gloria. *Are We Alone? Scientists Search for Life in Space.* Washington, D.C.: National Geographic Society, 2004.

Vogt, Gregory. *Our Universe: Exploring Space.* Austin, TX: Raintree-Steck Vaughn, 2001.

Books

Nardo, Don. *Extraterrestrial Life*. Farmington Hills, MI: Lucent Books, 2004.

Sherman, Joseph. *Deep Space Observation Satellites*. New York: Rosen Publishing Group, 2003.

Skurzynski, Gloria. *Are We Alone? Scientists Search for Life in Space*. Washington, D.C.: National Geographic Society, 2004.

Websites

Astrobiology at NASA
   http://astrobiology.arc.nasa.gov

Astrobiology Magazine
   www.astrobio.net

The Astrobiology Web
   www.astrobiology.com

NASA Astrobiology Institute
   http://nai.arc.nasa.gov

ABOUT THE AUTHOR

MARY FIRESTONE grew up in North Dakota. She lives in St. Paul, Minnesota, with her 11-year-old son, Adam, their pet beagle, Charlie, and their cat, Rigley. She has a bachelor's degree in music from the University of Colorado at Boulder and a master's degree in writing from Hamline University. When she isn't writing articles for magazines and newspapers and books for children, she enjoys gardening and spending time with her son.